Nature Center

Julie Murray

Abdo Kids Junior
is an Imprint of Abdo Kids
abdobooks.com

Abdo
FIELD TRIPS
Kids

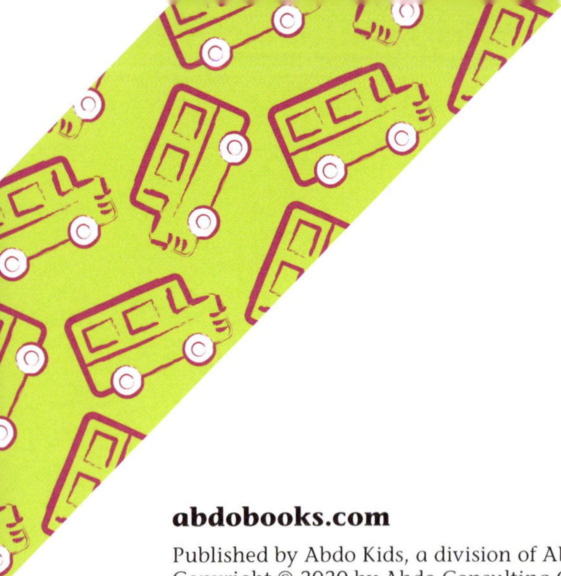

abdobooks.com

Published by Abdo Kids, a division of ABDO, P.O. Box 398166, Minneapolis, Minnesota 55439.

Abdo Kids Junior™ is a trademark and logo of Abdo Kids.

Printed in the United States of America, North Mankato, Minnesota.

102019

012020

Photo Credits: Alamy, Getty Images, iStock, Media Bakery, Shutterstock

Production Contributors: Teddy Borth, Jennie Forsberg, Grace Hansen

Design Contributors: Christina Doffing, Candice Keimig, Dorothy Toth

Library of Congress Control Number: 2019941213

Publisher's Cataloging-in-Publication Data

Names: Murray, Julie, author.

Title: Nature center / by Julie Murray

Description: Minneapolis, Minnesota : Abdo Kids, 2020 | Series: Field trips | Includes online resources and index.

Identifiers: ISBN 9781532188749 (lib. bdg.) | ISBN 9781532189234 (ebook) | ISBN 9781098200213 (Read-to-Me ebook)

Subjects: LCSH: Nature centers--Juvenile literature. | Nature study--Juvenile literature. | Ecology--Juvenile literature. | School field trips--Juvenile literature.

Classification: DDC 371.384--dc23

Table of Contents

Nature Center

It's field trip day. The class is going to the nature center.

It is a place to **explore**. It is a place to learn.

Bob sees the fish. He watches them swim.

Sofia walks on the path.

She looks at all the trees.

Sid learns about eagles.

The class holds a snake.

It is big!

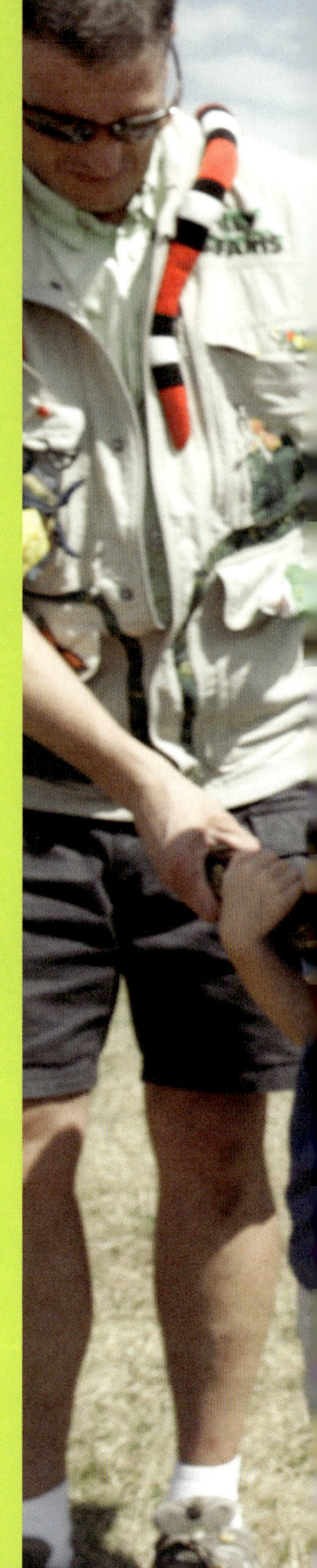

Addy builds a fort.

She uses sticks.

Mia is at the **pond**.

She sees a frog.

Have you been to
a nature center?

Life Cycle of Grasshopper
1 2 3 4 5 6 7 8 9

Nature Center Activities

build a fort

explore nature

go for a walk

learn about animals

Glossary

explore
to move through in order to learn.

pond
a small body of still water.

Index